IC 9518 ▪ *Information Circular*

Noise Control in Underground Metal Mining

by Efrem R. Reeves[1], Robert F. Randolph[2], David S. Yantek[3], and
J. Shawn Peterson[4]

DEPARTMENT OF HEALTH AND HUMAN SERVICES
Centers for Disease Control and Prevention
National Institute for Occupational Safety and Health

---

1 Acoustical Engineer, U.S. Army Aeromedical Research Laboratory, Fort Rucker, AL
2 Manager, Hearing Interventions Team, Pittsburgh Research Laboratory, National Institute for Occupational Safety and Health, Pittsburgh, PA
3 Research Engineer (Mechanical Engineer), Pittsburgh Research Laboratory, National Institute for Occupational Safety and Health, Pittsburgh, PA
4 Research Engineer (General Engineer), Pittsburgh Research Laboratory, National Institute for Occupational Safety and Health, Pittsburgh, PA

This document is in the public domain and may be freely copied or reprinted.

Disclaimer

Mention of any company or product does not constitute endorsement by the National Institute for Occupational Safety and Health (NIOSH). In addition, citations to Web sites external to NIOSH do not constitute NIOSH endorsement of the sponsoring organizations or their programs or products. Furthermore, NIOSH is not responsible for the content of these Web sites.

Ordering Information

To receive documents or other information about occupational safety and health topics, contact NIOSH at

Telephone: 1–800–CDC–INFO (1–800–232–4636)
TTY: 1–888–232–6348
E-mail: cdcinfo@cdc.gov

or visit the NIOSH Web site at www.cdc.gov/niosh.

For a monthly update on news at NIOSH, subscribe to NIOSH eNews by visiting www.cdc.gov/niosh/eNews.

DHHS (NIOSH) Publication No. 2010–111

December 2009

Contents

Acronyms and Abbreviations v

Acknowledgments vi

1 Introduction 1
Noise-Induced Hearing Loss 1
Three Variables of Noise Exposure 2
Noise Problem Analysis 2
Noise Dosimeters 3
Sound Levels 4
The Role of Engineering Noise Controls in Reducing NIHL 5

2 Noise Control 7
Hierarchy of Noise Control 7
Barriers and Sound-Absorbing Materials 7
Field Studies 8

3 Evaluating Noise Controls for Haul Trucks 11
Absorptive Material 11
Partial Engine Enclosures 15
Discussion 16

4 Evaluating Noise Controls for Load-Haul-Dumps 21
Engine Enclosure 1 21
Engine Enclosure 2 24
Enclosed Operator Cab 28
Discussion 30

5 Evaluating Noise Controls for Jumbo Drills and Bolters 33
Covers for Electric-Motor-Powered Hydraulic Pumps 33
Absorptive Material in Canopy 35
Absorptive Material on Sides of the Cab Around Operator Area 36
Absorptive Material in Lower Front of Cab 37
Windshields 37
Discussion 40

6 Summary and Guidelines 43
Haul Trucks 43

    Load-Haul-Dumps (LHDs) 43
    Jumbo Drills and Bolters. 44
    Lessons Learned. 45

References 47

Appendix A: What is Noise?. 49
    The Physics of Sound. 49
    Decibels. 51
    Average Sound Levels 52

Appendix B: Noise Exposure as a Health Hazard 55
    Effects of Noise. 55
    Dose Accumulation 55

Appendix C: Sound Measurement 57
    Sound Level Meters. 57
    Frequency Bands 61
    Decibel Addition 61

Acronyms and Abbreviations

ANSI	American National Standards Institute
CFR	Code of Federal Regulations
dB	decibels
dB(A)	decibels, A-weighted
f	frequency
Hz	Hertz
L_{eq}	equivalent continuous sound level
L_{avg}	average sound level
LHD	load-haul-dump
m	meters
MSHA	Mine Safety and Health Administration
NIOSH	National Institute for Occupational Safety and Health
NIHL	noise-induced hearing loss
Pa	Pascals
PEL	permissible exposure level
psi	pounds per square inch
REL	recommended exposure limit
rms	root mean square
SLM	sound level meter
TWA_8	time-weighted average (8-hour)
λ	wavelength

ACKNOWLEDGEMENTS

The authors are indebted to the following colleagues who participated in data collection, analysis, and reviews of early versions of this report: Daniel R. Babich, Michael DiMartino, Thomas M. Durr, Gerald L. Finfinger, Patrick J. Hintz, Peter G. Kovalchik, R.J. Matetic, Timothy J. Matty, Alex D. Prokop, and Adam K. Smith.

1. Introduction

1.1 Noise-Induced Hearing Loss

Noise-induced hearing loss (NIHL) is the most common occupational illness in the United States, with 30 million workers exposed to excessive noise levels [NIOSH 1996] every day. Of particular concern is the mining industry; which has the highest prevalence of hazardous noise exposure of any major industry sector [Tak et al. 2009] and is second only to the railroad industry in prevalence of workers reporting hearing difficulty [Tak and Calvert 2008].

This document is for operators, safety personnel, and mechanics in the mining industry who are not specialists in noise control engineering or acoustics. Evaluations of successful and unsuccessful attempts at controlling noise on several large, underground metal mining machines are detailed to illustrate the basic principles of noise control. Once personnel understand the guidelines and principles of noise control, they will be able to

- evaluate the extent of a noise problem;
- determine the best approach to the problem; and
- apply the most appropriate solution.

Because of the insidious nature of NIHL, it can go unnoticed until a considerable loss of hearing has occurred. In some cases, diagnosis is delayed because an exposed individual claims to have become accustomed to the noise. In reality, that person may have already suffered irreversible hearing loss.

Humans can hear sounds in the frequency range from about 20 to 20,000 Hertz (Hz). Within this range, NIHL usually begins in the frequency region around 4,000–6,000 Hz, the upper levels of the speech region. The first noticeable symptoms include difficulty understanding higher pitched voices, such as the voices of females and children, and difficulty understanding certain consonant sounds, which are primarily high frequency in nature.

The extent of NIHL varies depending on the level and duration of noise exposure and on an individual's susceptibility; despite having similar noise exposure, individuals can experience differing degrees of hearing loss, or none at all.

NIHL is almost always preventable. To reduce or eliminate the possibility of NIHL, an individual's noise environment must be analyzed and appropriate action taken to reduce noise exposure.

1.2 Three Variables of Noise Exposure

The three elemental components to consider when devising an engineering noise control are source, path, and receiver, which interact with each other to produce a unique situation for a given environment; the same source can yield different sound levels when the path or the location of the receiver is changed. Engineering noise controls can be implemented to reduce the amount of sound energy generated by the noise *source* and to divert the flow of sound energy from the propagation *path*, all with the aim of protecting the *receiver* (worker) from being exposed to high levels of sound energy.

1.3 Noise Problem Analysis

The Mine Safety and Health Administration (MSHA) established noise exposure limits for mine workers, and these are given in 30 CFR Part 62 (the *noise rule*). MSHA regulations state that an individual's occupational noise exposure should not exceed an 8-hour time-weighted average (TWA_8) of 90 dB(A). This level is defined as the permissible exposure level (PEL). Higher noise exposures are permitted for shorter periods of time as shown in Table 2. The noise rule applies to the sound received by a worker, not to sound emitted by a machine or process.

Table 1. Permissible time allowed for a given noise exposure

Exposure duration (hours)	Sound level (dB[A])
8	90
6	92
4	95
3	97
2	100
1	105
0.5	110
0.25	115

Source: 30 CFR Part 62

The noise rule stipulates that noise exposures resulting in the MSHA action level (AL) TWA_8 of 85 dB(A) or more require that hearing protection is provided to the exposed miner. Exposures resulting in a TWA_8 of 105 dB(A) or more require dual hearing protection. In practice, this usually means that an earmuff-type hearing protector must be worn over an earplug-type hearing protector. Dual hearing protection is required in addition to the actions required for enforcing the PEL. Further, the noise rule states that no miner can be exposed to sound levels exceeding 115 dB(A) for any amount of time.

Under the MSHA noise rule for the PEL, sound levels below 90 dB(A) do not contribute to the calculation of partial noise doses (Appendix B). In other words, a miner could be exposed to a sound level of 89 dB(A) for a full shift and—from a regulatory point of view—receive zero noise dose. This does not mean, however, that this individual has zero risk of receiving hearing damage. The NIOSH recommended exposure limit (REL) is a TWA_8 of 85 dB(A). Noise levels exceeding this REL are considered hazardous by NIOSH.

1.4 Noise Dosimeters

To determine the amount of noise workers are exposed to during the course of their day, workers can wear noise dosimeters. A dosimeter is designed to be worn on a person during all or part of a work shift, and it measures and stores sound levels and computes total noise exposure. Dosimeters are especially useful in environments where the noise levels are variable or intermittent or when workers move to and from different areas of a plant or mine during the course of a work shift. If sound levels are constant and the worker does not move, a sound level meter (SLM) can also be used to assess exposure. The procedure for using SLMs to measure noise and assess exposure is detailed in Appendix C.

Dosimeters and SLMs incorporate filters—or weighting networks—that can be applied to affect the meter's sensitivity to desired sound frequencies. The weighting is performed according to accepted standards. The A-weighting network approximates human perception of the loudness of low level sounds (around 40 dB). It is the most widely used weighting network because it is a reasonable estimator of the risk of NIHL. Without weighting in place, the SLM would indicate the same sound pressure level for sound waves having the same amount of physical energy regardless of the sound's frequency. In reality, very low and high frequency sounds are less damaging than mid frequency sounds. so A-weighting de-emphasizes the extreme frequencies. In the test examples in the following sections, the A-weighted scale is used, resulting in A-weighted decibels symbolized by units of dB(A).*

A dosimeter must be calibrated before and after each measurement period with a calibrator that fits the specific type of microphone for the meter. The pre-measurement calibration is necessary to ensure the instrument is functioning properly prior to making measurements. The post-measurement calibration is especially important in a mining environment because the instrument is likely to be subjected to jolting and jarring during a work shift and because temperature or humidity extremes could affect the accuracy of the meter.

---

* See Appendix C for more details about weighting networks.

The microphone, the most fragile part of the instrument, is especially susceptible to damage. The documentation provided by the instrument manufacturer should give the valid operating ranges for humidity, pressure, and temperature. The SLM should be calibrated by a qualified laboratory at the interval recommended by the manufacturer, typically every 1-2 years. Their calibration should be traceable to the National Institute of Standards and Technology.

Proper placement of the dosimeter microphone is important. ANSI S12.19-1996, *Measurement of Occupational Noise Exposure*, specifies that the microphone should be located on the mid-top of the wearer's most noise-exposed shoulder. The microphone should be set approximately parallel to the plane of wearer's shoulder, and the cable should be routed and fastened such that it does not interfere with job performance or create a safety hazard. For miners, the best place to attach the dosimeter case is usually the miner's belt.

1.5 Sound Levels

When a weighting filter is applied in determining the level of a sound, the word *pressure* is dropped from the term *sound pressure level*. For context, a list of the A-weighted average sound levels of some common sounds are shown in Table 2.

Table 2. Typical sound levels of common sources

Sound source	Level (dB(A))
Threshold of pain	120
Rock concert	110
Subway train	100
Heavy truck or bus (15 meters away)	90
Urban traffic	80
Passing cars (15 meters away)	70
Normal conversation	60
Classroom	50
Suburban neighborhood at night	40
Bedroom at night	30
Broadcast studio	20
Rustling leaves	10
Threshold of hearing	0

Source: Cowan 1994

1.6 The Role of Engineering Noise Controls in Reducing NIHL

The mining industry recognizes how important engineering noise controls are in reducing noise exposure during underground operations. But, because of the relatively small market for mining equipment, manufacturers have limited incentives to develop less noisy machinery or more innovative noise controls. Also, the specialized equipment designs imposed by the sometimes-hostile mining environment has limited the transfer of noise control technologies from other industries.

Despite this lack of proven control technologies, mine operators work with what's available to try to create noise control solutions at the mine level. However, many operators install noise treatments without knowing how much noise reduction to expect or how much noise reduction is actually achieved after installation. In some cases, because of improper material selection, placement, or installation, the noise treatment reduces sound little—if any. In other cases, noise treatments are applied when the source sound level does not warrant treatment, thus wasting effort and resources. Unsuccessful noise controls cost the industry time and money, and they do nothing to decrease workers' risk of NIHL—though they give the false impression that the problem, if there is one, has been addressed.

2 Noise Control

2.1 Hierarchy of Noise Control

Three methods to reduce worker noise exposure are:

1. Implementing engineering noise controls to reduce noise at the source or at the worker
2. Using administrative controls to limit the amount of time workers spend in noisy environments
3. Wearing personal protective equipment, such as hearing protectors, to reduce the sound level entering the ears.

Using engineering noise controls is the most desirable option because they address noise sources directly. Administrative controls and hearing protectors are indirect interventions and are less easily monitored and therefore more readily circumvented.

2.2 Barriers and Sound-Absorbing Materials

A barrier is a solid obstacle that is at least somewhat impervious to sound and interrupts the direct path from the sound source to the receiver. The sound transmission loss (TL) of a material is a measure of its ability to block sound. To block sound most effectively, the barrier should be

- placed as close as possible to either the source or receiver;
- assembled to be as tall and wide as practical so it extends well beyond the direct source-receiver path; and
- constructed of a material that is solid and airtight.

Low frequency sounds are difficult to block with barriers because low frequency sounds pass directly through and bend around obstacles relatively easily. This is why the bass tones from a passing car stereo are audible even inside buildings. Mid to high frequency sounds, which often dominate a worker's noise exposure, cannot pass through or bend around barriers as easily as low frequency sounds. In general, adding mass to a barrier improves its ability to block noise. Another way to improve the TL of a barrier is to use multiple layers of material with each layer separated from the others using a compliant material such as foam. This method decouples the vibration of each layer from the other layers and, therefore, increases the TL.

Sound-absorbing treatments are usually made of porous materials that absorb incident sound energy and reduce the reverberation due to sound reflected from surfaces. Fiberglass and open-cell foam are often used for sound absorbers. A material's degree of sound absorption depends on its flow resistance and thickness, the way it is mounted, and the frequency of the incident sound. Thicker sound absorbing materials are needed to absorb low frequency sounds. For frequencies above about 1 kHz, 1-inch-thick sound absorbing material has sufficient sound absorption. Two-inch-thick sound-absorbing material has good absorption for frequencies above about 500 Hz. Protective facings on sound absorbing foam tend to improve the sound absorbing capabilities of the material at lower frequencies.

To improve the sound absorption of an installed material, the material can be mounted with an air space between it and the surface behind it. To achieve the best results, the material should be spaced one-quarter wavelength from the surface behind it. In this case, the wavelength is based on the lowest frequency of interest. In addition, the optimal absorption of a material occurs when the thickness is equal to one-quarter wavelength for the frequency of interest. Or,

$$t = \lambda/4 \qquad (1.1)$$

where t = material thickness, inches
λ = wavelength of sound for lowest frequency of interest, inches

Equation 1.1 shows that in some instances it can be impractical to install material with the optimal thickness or spacing to absorb low frequency sounds. For example, the optimal material thickness for noise at 1 kHz is roughly 3.5 inches and the optimal material thickness for noise at 500 Hz is about 7 inches. Knowing the frequency content of a noise problem enables one to select a sound-absorbing material that has sufficient absorption at the frequencies where the noise energy is greatest.

2.3 Field Studies

In its advice concerning controlling noise, MSHA emphasizes a reduction in daily noise exposure or dose. To accommodate this, noise controls should be evaluated in the environments in which they will be used. Initially, a sound level meter should be used to assess a noise control by measuring the sound level without and with the noise control. If the application of a noise control does not reduce sound levels, then this control is ineffective at reducing noise, assuming all other factors have remained constant. Once a control is applied that reduces sound levels, changes in noise exposure can be assessed through dosimetry.

Engineering noise controls can be difficult and expensive to implement. For complex noise problems, it is best to consult with a professional noise control engineer before implementing a solution. However, some simple measures to reduce sound levels, including proper maintenance of mining equipment, are easy to implement. This document details results of testing of barriers and sound-absorbing materials used to control sound levels on and around machinery at several mine sites. To show the detail of the measurements, the sound levels are reported to the nearest 0.1 dB. However, for general use, rounding to the nearest whole number is permissible.

3 Evaluating Noise Controls for Haul Trucks

Haul trucks at metal mines operate both underground and on the surface. Haul truck noise is a good example of the challenging problem of reducing noise in both reverberant and non-reverberant environments.

3.1 Absorptive Material

Several of the tested haul trucks had 0.75-inch-thick, vinyl-covered material installed in the area in front of the operator. It is not known if this material was a sound absorbing material or if it was only padding. Figure 1 shows an example of these haul trucks. The material was attached with Velcro for easy removal (Figure 2).

NIOSH researchers measured sound levels at the operator position above ground at low and high idle, with and without the vinyl-covered material in place. Table 3 gives the results. The data for haul truck 1 confirm that the vinyl-covered material had little effect on sound levels at the operator's position. The results for haul truck 2 show that the sound level at low idle was higher with the vinyl-covered material than without it, possibly due to fluctuations in engine output between the tests. At high idle, the vinyl-covered material had no measurable effect on the sound level.

Chapter 3 | Evaluating Noise Controls for Haul Trucks

Figure 1. Haul truck with vinyl-covered material installed in the area in front of the operator.

Figure 2. Haul truck without vinyl-covered material.

Table 3. Sound level at the haul truck operator's position, surface measurement

	Haul truck 1 without material (dB[A])	Haul truck 1 with material (dB[A])	Haul truck 1 noise reduction (dB[A])	Haul truck 2 without material (dB[A])	Haul truck 2 with material (dB[A])	Haul truck 2 noise reduction (dB[A])
Low idle	81.9	81.7	0.2	85.3	86.5	-1.2
High idle	101.3	100.6	0.7	101.3	101.3	0.0

The third haul truck had the same 0.75-inch-thick, vinyl-covered material in the area in front of the operator as the previous two haul trucks. However, haul truck 3 also had vinyl-covered material attached to the underside of the canopy (see Figures 3 and 4). NIOSH researchers compared sound levels measured at the surface with those measured underground under the same conditions: at the operator position at low and high idle, with and without the vinyl-covered material in place.

Figure 3. Haul truck with vinyl-covered material installed in canopy above the operator.

Chapter 3 | Evaluating Noise Controls for Haul Trucks

Figure 4. Haul truck with vinyl-covered material removed.

Table 4 shows the effect of having vinyl-covered material in the canopy by comparing the measurement with all material installed to the measurement taken without material in the canopy or in front of the operator. Table 4 also shows the effect of using all the material in the underground environment. At high idle, using the vinyl-covered material only on the underside of the canopy reduced the sound level by 0.6 dB(A) whereas installing all the material reduced the sound level by 1.0 dB(A).

Table 4. Sound level at the haul truck operator's position, underground measurement

	Without vinyl-covered material (dB[A])	With vinyl-covered material in canopy only (dB[A])	With all vinyl-covered material (dB[A])	Noise reduction with vinyl-covered material in canopy (dB[A])	Noise reduction with all vinyl-covered material (dB[A])
Low idle	83.2	82.4	82.6	0.8	0.6
High idle	100.6	100.0	99.6	0.6	1.0

Table 5 shows the sound levels for haul truck 3 measured above ground at the operator's position. Comparing the data from Table 5 with that of Table 4, the underground environment increased the sound level at the operator's ear by 2.4 dB(A) at low idle and 1.5 dB(A) at high idle.

Table 5. Sound level for haul truck 3 at the operator's position, surface measurement

	Full padding (dB[A])
Low idle	80.2
High idle	98.1

3.2 Partial Engine Enclosures

Engine enclosures are used to contain engine noise. Sound-absorbing material can be used to line engine enclosures to absorb noise within the enclosure. This can reduce the sound level emitted from the enclosure. The actual amount of noise reduction achieved depends on many factors. To contain engine noise, the enclosure must be made from a material with a high TL. Adequate space is needed between the engine and its enclosure to allow proper flow of cooling air. If the space between the engine and enclosure is insufficient, the cooling fan will not be able to efficiently move air and the noise due to the fan may increase substantially. Gaps in an enclosure greatly reduce its ability to contain noise.

To test how effective partial engine enclosures are at reducing sound levels on haul trucks, sound levels were measured underground at the haul truck operator's position. Data were collected on eight different haul trucks at high idle. Each haul truck had a partial engine enclosure similar to the one shown in Figure 5, fashioned from a piece of 0.5-inch-thick rubber that NIOSH researchers believed to be used conveyer belt material. Figure 6 shows the engine without the partial enclosure. Measurements were made with and without the barrier in place. The results showed that the barrier reduced the sound level reaching the operator by about 1 dB(A).

Chapter 3 Evaluating Noise Controls for Haul Trucks

Figure 5. Partial engine enclosure.

Figure 6. Open engine compartment.

3.3 Discussion

3.3.1 Absorptive Materials

The results of this testing demonstrate that using absorptive materials is not a cure-all fix for haul truck noise. Often, the low cost and relative ease of application of some absorptive materials leads inexperienced personnel to apply them as a quick fix. However, before installation the situation must be analyzed to determine whether they will be effective. Without this initial assessment, the time and expense of applying the materials may be completely wasted.

For example, Figure 7 shows a haul truck with an open cab. The arrows represent sound waves and the blue area in the canopy of the haul truck represents sound-absorbing material. For the material to be effective, the majority of the sound reaching the operator must be due to reflections from the underside of the canopy. In this example, the operator would be in the direct path of the sound waves from the engine compartment. In addition, sound waves reflected from the rib have a direct path to the operator. Only a fraction of the sound waves reaching the operator are reflected from the underside of the canopy. So, in this case, the material would have little to no effect.

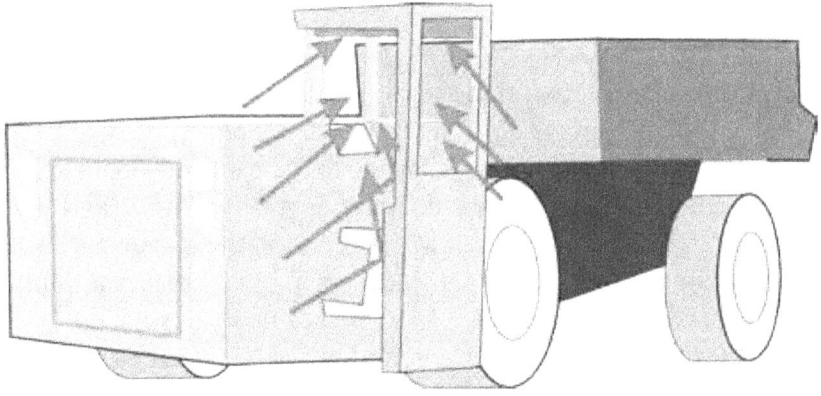

Figure 7. Haul truck with sound-absorbing material installed in canopy and depiction of how sound may enter the operator area, reaching operator before padding.

The test results showed that the underground environment increased sound levels at both low and high idle. NIOSH researchers attribute this increase to the reverberation of sound that occurs in enclosed spaces. The amount of increase also depended on whether the machine was running at low idle or high idle. This is due to the different frequency content associated with the noise emitted at high and low idle and how each of these is affected by the mine environment. Since the environment and operating conditions can have a significant impact on equipment noise, controls should be assessed in the environment where they are used under all operating conditions.

3.3.2 Partial Enclosures

When use of a full enclosure to control noise is impractical, testing has shown that a properly designed partial enclosure or barrier can provide some level of sound reduction. When a partial enclosure is used, it is critical to block the line of sight to the engine as much as possible. Adjacent sections of material should be overlapped to prevent sound waves from escaping the engine compartment.

Engine enclosures cannot be sealed completely due to requirements for cooling air. Airflow paths should be designed such that the air must follow a maze-like path. The surfaces of the airflow path should be lined with sound-absorbing material. As noise travels down the flow path, the sound-absorbing material will absorb some of the sound energy.

3.3.3 Sealing Gaps

An often overlooked noise control measure is sealing gaps. A hole or gap in an enclosure, even if small, can greatly compromise noise reduction. Gaps provide a direct path for sound to travel from the engine to the haul truck operator. Sealing gaps reduces the noise exposure of the operator. Figure 8 shows a large gap around the perimeter of the instrument panel, which is part of the engine enclosure. Sealing the gaps around the instrument panel, as shown in Figure 9, can significantly reduce the operator's noise exposure. Sound-absorbing foam should not be used to seal gaps. Due to its open cell nature, sound-absorbing foam is not very good at blocking noise. When sealing gaps, closed cell foam should be used instead.

Figure 8. Direct openings to the engine compartment around instrument panel.

Figure 9. Gaps around instrument panel sealed with foam.

4 Evaluating Noise Controls for Load-Haul-Dumps

Load-haul-dump vehicles (LHD) are widely used in underground metal mines for moving material. NIOSH researchers assessed several noise controls applied in the engine compartment and the operator's cab.

4.1 Engine Enclosure 1

LHD1 is shown in Figure 10. The engineering noise controls on the machine consisted of a partial engine enclosure and sound-absorbing material in the engine compartment and in the cab. The engine enclosure on the left (operator) side of the machine was composed of 0.125-inch-thick steel panels insulated with 1.5-inch-thick fiberglass sound-absorbing material. Engine areas not covered by steel were covered with 0.25-inch-thick rubber that NIOSH researchers believed to be used conveyer belt material. The left side of the engine was completely enclosed as seen in Figure 11.

Figure 10. LHD1.

Figure 11. Left side of the engine enclosure.

Figure 12 shows the front and back of the steel panels that covered the left side of the engine. The right side of the engine enclosure, shown in Figure 13, consisted of 0.25-inch-thick rubber that did not completely cover the engine compartment opening.

Figure 12. Front and back of the steel panels insulated with 1.5-inch-thick fiberglass.

Figure 13. LHD1 partial engine enclosure, right side.

Sound levels at the operator's position were measured both on the surface and underground with the engine at high idle. Measurements were made with both sides of the engine enclosure on or off, and then with the right side off and the left side either on or off. Table 6 gives the results. The measurements on the surface show that the full enclosure reduced the sound level by about 1 dB(A). Because these LHDs are primarily used underground, the underground results are more important. Comparing the surface and underground measurements, the underground environment adds about 3 to 4 dB(A) to the sound level at the operator's ear. The table shows the application of all controls in the underground environment resulted in an attenuation of 1.5 dB(A) at the operator position.

Table 6. Sound level for LHD1 at the operator's position, high idle

	Surface measurement			Underground measurement		
	Without enclosure (dB[A])	With enclosure (dB[A])	Noise reduction (dB[A])	Without enclosure (dB[A])	With enclosure (dB[A])	Noise reduction (dB[A])
Both sides on or off	91.9	90.8	1.1	95.8	94.3	1.5
Right side off, Left side on or off	91.6	90.8	0.8	95.1	94.3	0.8

4.2 Engine Enclosure 2

LHD2 is shown in Figure 14. The noise controls on the machine consisted of a partial engine enclosure, absorptive material in the engine compartment, and professionally installed noise control material in the open cab. The top portion of the engine enclosure was composed of 0.375-inch-thick steel panels lined with quilted fiberglass sound-absorbing material. The left side of the engine was almost completely enclosed, as shown in Figure 15, by 0.125-inch-thick steel panels with quilted sound absorption. However, the picture shows several gaps around the perimeter of the steel panels. Figure 16 shows the inside of the steel panels covered with absorptive material. The right side of the engine compartment was not enclosed as shown in Figure 17.

Sound levels were measured at the operator's position with the engine at low and high idle with the noise controls described above. At low idle, the sound levels were about 75 dB(A) on the surface and slightly less than 80 dB(A) underground. With the engine at low idle, the sound level was measured underground with the machine in reverse and the back-up alarm sounding. In this case, the sound level was 95.4 dB(A). Table 7 shows the A-weighted sound levels measured at the

Figure 14. LHD2.

Figure 15. Left side partial engine enclosure.

Chapter 4 — Evaluating Noise Controls for Load-Haul-Dumps

Figure 16. Quilted absorber inside the engine compartment.

Figure 17. Right side and top of engine compartment.

operator position at high idle on the surface and underground. On the surface, the engine enclosure offers 1.5 dB(A) of noise reduction. However, examination of the underground results shows a reduction of only 0.1 dB(A). This is most likely due to noise from the right side of the engine reflecting from the rib and back to the operator station. Figure 17 shows that the right side of the engine is completely exposed.

Table 7. Sound level for LHD2 at the operator's position, high idle

	Sound level without enclosure (dB[A])	Sound level with enclosure (dB[A])	Noise reduction (dB[A])
Surface	89.4	87.9	1.5
Underground	92.2	92.1	0.1

Figure 18 shows the one-third-octave-band sound pressure level spectrum measured underground at the operator's position with the engine at high idle without the partial enclosure. The sound level was the same as that of LHD2, 92.2 dB(A), as shown in the column labeled "Without enclosure/Underground" of Table 7. The figure shows that most of the sound energy is below 500 Hz. To effectively absorb

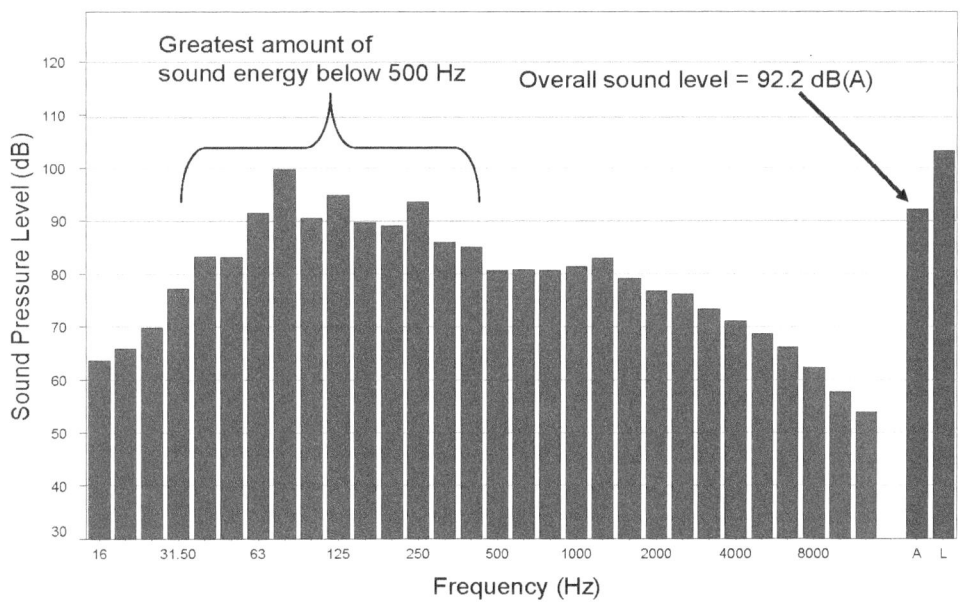

Figure 18. In-cab one-third-octave band spectrum for LHD2 at high idle with engine compartment open.

low frequency sounds, a thick material (4–7 inches) or a thin material backed by an airspace is required [Moulder 1998]. Unfortunately, either solution would be difficult to implement in the limited space of the engine compartment.

It is interesting to note that the 80 Hz one-third-octave band is the highest. Noise at the engine firing frequency is contained in this band. Perhaps a larger muffler would have reduced the sound level of the machine. The presence of significant low frequency energy points to the cooling fan as a likely noise source. Neither exhaust noise at the engine firing frequency nor cooling fan noise would be significantly affected by noise controls applied to the engine enclosure.

4.3 Enclosed Operator Cab

Figure 19 shows a 6-yard LHD similar to LHD3. (The machine in the figure is not the actual machine evaluated.) The noise control installed on the evaluated machine consisted of a glass-enclosed cab lined with vinyl-faced, 1-inch-thick open cell foam with an attached barrier material. The material used in the cab is shown in Figure 20.

To enclose the cab, shown in Figure 21, window-frame and upper door modifications were required for glass panes to be installed within the frames. All of the modifications were cleared with the manufacturer to ensure that the integrity of the falling object protective structure was not adversely affected.

Figure 19. LHD3.

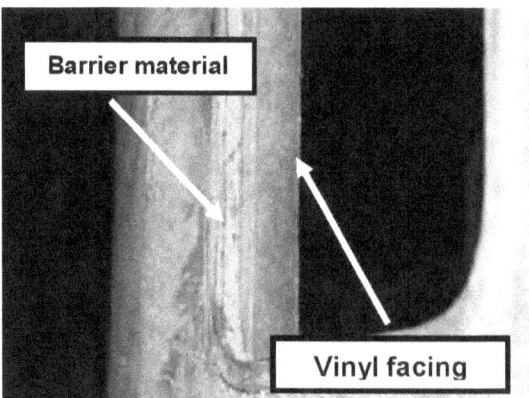

Figure 20. Open cell foam used for in-cab sound absorption.

Figure 21. Enclosed cab with glass in place.

With the machine underground, NIOSH researchers simultaneously measured sound levels inside and outside of the cab with the engine at high idle. In addition, sound levels were measured above, in front, to the right, and to the rear of the cab with and without the glass panels installed.

Table 8 shows the measurement results. The column labeled "Interior" is the average of four measurements taken inside the cab. The "Exterior" column is the average of the four exterior measurements and can be used as a reference to judge the noise reduction provided by the cab. The exterior measurements were also used to ensure that the sound level generated by the LHD did not vary much during the course of the measurement period. The results indicate that the completely enclosed cab produced greater than 20 dB(A) of noise reduction. Even without the front window installed (row 4), more than 10 dB(A) of reduction was achieved.

Table 8. Sound level for LHD3 at the operator's position, high idle, underground measurement

	Exterior (dB[A])	Interior (dB[A])
All windows open	99.9	96.9
Left window closed	98.2	93.8
Back and left windows closed	98.4	92.9
Back, left, and right windows closed	99.9	89.1
All windows closed	100.3	77.7

4.4 Discussion

Partial engine enclosures with openings of any size greatly compromise the noise reduction capability of the enclosure. This is especially true underground, where sound initially directed away from an operator can strike the walls and reflect back to the operator. To be effective at reducing the sound levels reaching the operator, enclosures must be designed to minimize holes and gaps, especially those with line of sight between the noise source and the operator.

The most effective noise-reducing enclosures are airtight. However, an airtight enclosure for a source that requires ventilation, such as an engine, is impractical because it could lead to overheating and engine damage. The only openings in the engine compartment should be those to allow cooling air into and out of the cooling package. For an LHD, if solid panels cannot be used for the engine enclosure, a partial enclosure that incorporates overlapping materials or baffles, similar to that suggested for haul trucks, should be used. Using a partial engine enclosure will decrease the sound levels compared to an open engine compartment. However, an engine compartment with solid panels is the best approach.

As a rule of thumb, enclosures should be lined with sound-absorbing material to reduce build up of reverberant noise within the enclosure. Full coverage of all surfaces is not completely necessary as the effect of adding sound-absorbing material decreases as more and more of the surfaces are covered. The best approach to develop an enclosure is to first eliminate any gaps or leaks and then to add sound-absorbing material inside.

For LHD2, the lined partial engine enclosure reduced the sound level by 1.5 dB(A) above ground. However, in the underground environment, the sound levels were not affected (refer to Table 7). This is probably due to an increase in the contribution of cooling fan and exhaust noise to the sound level at the operator's ear in the underground environment. Underground, fan noise and exhaust noise can reflect from the rib to the operator. In addition, the underground environment may have amplified the exhaust tones corresponding to the engine firing rate.

A fully enclosed environmental cab can provide 20 dB(A) or more of noise reduction. Besides providing protection and comfort for the operator, environmental cabs are designed to reduce exposures to occupational hazards such as dust and noise. When installing a retrofit cab, it is wise to contact the original equipment manufacturer to ensure that the integrity of the falling object protective structure is not compromised. In addition, once the cab is enclosed, a climate control system should be installed to ensure the safety and comfort of the equipment operator.

The results of testing done with LHD3 indicate that the completely enclosed cab reduced noise by more than 20 dB(A). Sound levels were reduced more than 10 dB(A) even with the back window removed. This shows that if completely enclosing an operator's cab is not feasible, a properly designed 3/4-cab enclosure can provide a substantial noise reduction. The resulting noise reduction will depend on the location of noise sources on the machine relative to the open area of the 3/4-cab. If there are no significant noise sources near the open area, the partial cab can still provide a substantial noise reduction. However, if a noise source has line of sight with the operator due to the exclusion of a side, the partial cab will not be effective.

5 Evaluating Noise Controls for Jumbo Drills and Bolters

Noise controls for jumbo drills and bolters consisted of several motor covers or barriers, treatments applied inside the cabs, and different cab windshield designs.

5.1 Covers for Electric-Motor-Powered Hydraulic Pumps

Ten machines were tested: five roof bolters and five jumbo face drills. All of the tested face drills and bolters were equipped with at least one electric motor used to drive hydraulic pumps. The dual-boom face drills were equipped with two electric motors used to drive hydraulic pumps. The motors were directly behind the operator area as shown in Figure 22. Five of the tested machines—two roof bolters and three jumbo drills—had noise controls installed around the electric motor and hydraulic pumps. All of the reported measurements were made underground at the operator's ear position with only the electric motors operating.

Figure 22. Dual-boom jumbo face drill.

Table 9 shows the sound levels without and with noise controls applied to the electric-motor-powered hydraulic pumps. Several of the tested controls are shown in Figures 23–25. It should be noted that the sound levels generated with only the electric motors on were less than 85 dB(A). Sound levels during drilling and bolting can exceed 100 dB(A). Therefore, the noise due to the electric-motor-powered hydraulic pumps is insignificant in terms of the operator's dose. The data show the motor enclosures built from barrier-type materials reduced the sound level by about 2 dB(A). However, the enclosures built from absorptive material reduced the sound level less than one-half dB(A). Sound-absorbing materials do not usually provide much TL.

Table 9. Sound level for jumbo drills and bolters at the operator's position, underground

Motors	Noise control	Without control (dB[A])	With control (dB[A])	Noise reduction (dB[A])
Bolter 1	0.25-inch-thick heavy conveyor belt	84.9	83.2	1.7
Bolter 2	1.5-inch-thick fiberglass blanket	77.3	76.9	0.4
Face drill 1	0.5-inch-thick heavy conveyor belt	79.4	77.2	2.2
Face drill 2	1.5-inch-thick quilted fiberglass absorptive material	79.9	79.5	0.4
Face drill 3	0.25-inch-thick Plexiglas	84.3	81.9	2.4

Figure 23. Heavy conveyor belt barrier.

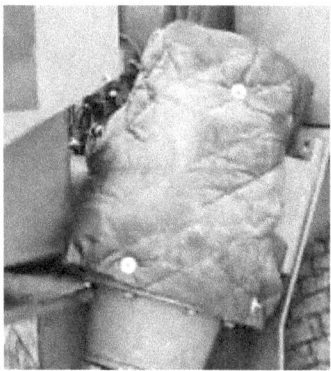

Figure 24. Fiberglass blanket barrier.

Figure 25. Plexiglas motor cover.

5.2 Absorptive Material in Canopy

Most of the tested machines had sound-absorbing material under the canopy. However, only three of the machines had the material installed in such a way that it could be easily removed and replaced so that NIOSH researchers could directly measure its effectiveness at reducing noise. In all of the cases, the absorptive material was a 1-inch-thick quilted fiberglass blanket. Figure 26 shows the material being removed for testing.

Figure 26. One-inch-thick quilted fiberglass blanket being removed for testing.

Table 10 shows the sound levels without and with the material and the resulting noise reduction achieved by applying it to the canopy. The face drill measurements were taken underground during the drilling cycle, and the bolter results were measured above ground with the percussive hammer operating. Face drill 2 had a removable windshield, so the effect of the absorptive material in the canopy was measured with and without the windshield. The data show the sound-absorbing material did not significantly change the sound levels at the operator's position in this case.

Table 10. Sound level of jumbo drills and bolters at the operator's position

Motors	Without absorptive material (dB[A])	With absorptive material (dB[A])	Noise reduction (dB[A])
Bolter 2	97.4	97.3	0.1
Face drill 1	99.1	99.3	-0.2
Face drill 2 (with windshield)	99.6	99.6	0.0
Face drill 2 (without windshield)	100.3	100.1	0.2

5.3 Absorptive Material on Sides of the Cab Around Operator Area

One bolter and one face drill had a removable 1-inch-thick quilted fiberglass blanket around the operator's area. For bolter 1, measurements were performed underground with the windshield in place while drilling and bolting. The operator's area of bolter 1 is shown in Figure 27. For face drill 2, the sound levels were measured while only the electric-motor-powered hydraulic pumps were operating.

Figure 27. Quilted fiberglass material in the operator's area of bolter 1.

Table 11 shows the sound levels with and without the absorptive material placed around the operator and the resulting noise reduction. The data indicate that the absorption around the operator has essentially no effect on the sound level during the drilling process. During the bolting process, the measured sound level at the operator's ear was 0.3 dB(A) higher with the material in place. However, this difference is most likely due to changes in the noise produced at the drill steel, not due to the installation of the sound-absorbing material.

Table 11. Sound level of jumbo drills and bolters at the operator's position, absorptive material around operator

	Without quilted material around operator area (dB[A])	With quilted material around operator area (dB[A])	Noise reduction (dB[A])
Bolter 1 (drilling)	97.5	97.6	-0.1
Bolter 1 (bolting)	98.4	98.7	-0.3
Face drill 2 (motor)	78.1	77.2	0.9

5.4 Absorptive Material in Lower Front of Cab

Figure 28 shows a 1-inch-thick quilted fiberglass blanket applied to the lower front of the operator area on bolter 2. This material is located where the operator's knees would be positioned while operating the drill boom.

Figure 28. Quilted fiberglass material in the lower front of the operator's area of bolter 2.

Table 12 shows the effect that the absorptive material placed in this position had on sound levels measured at the operator's ear during drilling and bolting. The table shows the levels were virtually unchanged in each case. This is not surprising. Most of the drilling and bolting noise probably reaches the operator by bending around the windshield or by first reflecting off the rib. The noise reflected from the front lower area to the operator is most likely minimal.

Table 12. Sound level of bolter 2 at the operator's position, absorptive material in lower front of cab

	Without absorption (dB[A])	With absorption (dB[A])	Noise reduction (dB[A])
Bolter 2 (drilling)	98.1	97.9	0.2
Bolter 2 (bolting)	99.9	99.9	0.0

5.5 Windshields

The most common noise control installed on the tested face drills and bolters was a windshield. The amount of noise reduction achieved varied greatly depending on how the windshield was designed. Several examples of windshields are shown in Figures 29–31.

Figure 29. Windshield installed on bolter 2.

Figure 30. Windshield installed on bolter 3.

Figure 31. Windshield installed on bolter 5.

Most of the windshields were designed to be flipped up into the canopy. This feature allowed the operator an unobstructed view while operating and tramming the machine. The windshield on bolter 2 had gaps between sections that were arranged vertically and did not wrap around the operator station (see Figure 29). The windshield of bolter 3 had no gaps between sections of windshield, and the windshield wrapped around the operator (see Figure 30). Bolter 5's windshield was continuous, but it did not wrap around the operator station. Strips of belting material had been installed on the sides of the operator station on bolter 5 in an effort to block noise (see Figure 31). Face drills 1 and 4 had wrap-around windshields without gaps between sections.

Table 13 shows the effect the windshields had on sound levels reaching the operator's ear. The greatest noise reductions were achieved for bolter 3, face drill 1, and face drill 4, all having wrap-around windshields with no gaps. The only difference between the windshields of bolters 2 and 5, was that bolter 2's windshield had gaps between panes and bolter 5's windshield was continuous. Bolter 5 experienced a 1-dB(A) greater noise reduction than bolter 2 when drilling.

Table 13. Sound level of bolters 2, 3, 4, and 5 at the operator's position

	Without windshield (dB[A])	With windshield (dB[A])	Noise reduction (dB[A])
Bolter 2 while drilling	98.5	97.9	0.6
Bolter 2 while bolting	101.2	99.9	1.3
Bolter 5 while drilling	100.6	99.0	1.6
Bolter 3 while drilling	99.2	96.0	3.2
Bolter 3 while bolting	105.7	102.5	3.2
Face drill 1	101.7	99.3	2.4
Face drill 2	100.1	99.6	0.5
Face drill 3	97.1	95.3	1.8
Face drill 4 with single boom	94.0	91.9	2.1
Face drill 4 with double boom	98.9	95.6	3.3
Face drill 5	101.9	100.6	1.3

5.6 Discussion

Covers for electric-motor-powered hydraulic pumps constructed of a heavy barrier material, such as conveyor belting, as opposed to an absorptive material such as fiberglass, produced the most substantial sound level reductions. However, on the tested machines the A-weighted sound levels created by the untreated motors were below 85 dB(A). Section 3.3.1 advised having the environment analyzed for noise levels prior to incurring the expense of noise treatments. If multiple noise sources generate sound levels of 85 dB(A) individually, it may be necessary to treat each of these sources to reduce the operator's noise exposure. For example, four 85-dB(A) noise sources operating together would result in a sound level of 91 dB(A). However, on a case-by-case basis, the contribution of each noise source to the operator's noise exposure should be determined before installing noise controls. With bolters and jumbo drills, the sound level due to drilling and bolting often reaches 100 dB(A), whereas the pumps generate a sound level less than 85 dB(A). Therefore, the noise exposure due to the electric-powered-hydraulic pumps is insignificant and, in this case, noise controls should not be applied to the pumps.

The application of fiberglass absorptive material to the canopy, seat area, and lower portion of the open cab had little to no effect on the sound level at the operator's ear during drilling and bolting. To be effective at reducing the sound level reaching the operator, sound-absorbing materials must be placed on surfaces that reflect sound toward the operator's hearing zone. Furthermore, a significant portion of the noise at the operator's ear must be due to noise reflected from these surfaces. If the majority of the noise at the operator's station arrives directly from the face or from reflections from the rib, treating the surfaces around the operator will have virtually no effect on the sound level at the operator's ear. For machines with open cabs, such as those installed on the face drills and roof bolters tested, absorptive materials will be of limited benefit.

For face drill 2, a reduction of nearly 1 dB(A) was achieved with absorptive material in the operator area with only the electric-motor-powered hydraulic pumps in operation. This reduction probably occurred because the noise from the pumps must reach the operator by an indirect path. Line of sight with the pumps is obscured by body of the machine. As this noise reflects off surfaces around the operator, the material around the operator reduces the noise. However, when the operator began drilling, the primary noise source became drilling noise. Since drilling noise reaches the operator mainly via a direct path, or by bending around the windshield, the absorbing material around the operator would have no effect.

In general, well-designed windshields were the most effective noise controls implemented on the drills and bolters because they block drilling and bolting noise from reaching the operator. Also, the noise generated by drilling and bolting is predominantly high frequency in nature. High frequency sounds are easier to block and absorb because of their shorter wavelengths. The windshields that had a gap between an upper and a lower pane of glass were the least effective at reducing sound levels because the gaps allow drilling and bolting noise to pass through.

The conveyor belt strips serving as a makeshift enclosure on bolter 5 were installed in an attempt to supplement the noise reduction due to the windshield. Because no sound level measurements were taken without the strips in place, the noise reduction they offer is unknown. NIOSH researchers assume they have little, if any, effect on sound levels reaching the operator's ear because of gaps between the strips. The strips should be overlapped a few inches to improve the noise reduction due to their use.

6 Summary and Guidelines

Engineering noise controls are the preferred solution to a noise problem because they address noise sources directly. Administrative controls and personal protective equipment should be explored as secondary solutions.

Basic noise controls include barriers and sound-absorbing materials. A barrier is a solid obstacle that is somewhat impervious to sound and that interrupts the direct path from the sound source to the receiver. For the best reduction in sound level, the barrier should be

- placed as close as possible to either the source or receiver;
- assembled to be as tall and wide as practical so it extends well beyond the direct source-receiver path; and
- constructed of a material that is solid and airtight.

Sound-absorbing treatments reduce reflections and the resulting echoes and reverberation. Usually, these materials are porous. Compared to high frequency sounds, low frequency sounds are more difficult to absorb with materials and to block with barriers. Therefore, it is important to know the frequency content for a particular noise problem.

The effectiveness of barriers and absorptive materials as noise controls on mining equipment was tested during field studies. Following are some of the key findings.

6.1 Haul Trucks

The use of absorptive materials in the operator's area of tested haul trucks had very little effect on sound levels underground. Sound level reductions were on the order of 1 dB(A). Most of the sound reaches the operator via the direct path from the noise source to the operator. In addition, noise reflects from the walls to the operator station. Open cabs allow the direct and reflected sound to enter the operator station. Therefore, a large reduction in sound levels from installing sound-absorbing material at the operator station is not expected.

6.2 Load-Haul-Dumps (LHDs)

A fully enclosed environmental cab can provide 20 dB(A) or more of noise reduction. If a fully enclosed cab is impractical, a partial cab can provide useful protection as long as the openings face away from the primary noise sources. A partial cab with three sides and a top was found to provide more than 10 dB(A) of noise reduction. Both full and partial cabs should have similar results on other underground equipment. When installing a retrofit cab, it is wise to contact the original equipment manufacturer to ensure that the integrity of the falling object protective structure (FOPS) is not compromised.

6.3 Jumbo Drills and Bolters

When applying noise control treatments, care should be taken to use the right product for the job. The 0.5-inch-thick rubber conveyor belt mats used to cover the electric-motor-powered hydraulic pumps on the jumbo drills and bolters were effective at reducing noise because the heavy rubber is a barrier material, which is the correct choice for the application. Rubber is usually not the best material to use for a barrier, but in this case it was effective. On bolter 2, the electric motor and hydraulic pumps were covered with sound-absorbing material. In this instance, the treatment had almost no effect on the noise from the electric motor and hydraulic pumps because sound-absorbing material makes a poor barrier. Sound-absorbing material is most effective when it is used at a reflective surface. The cover should have been constructed using a barrier lined with sound-absorbing material to surround the electric motor and hydraulic pumps.

Prior to developing noise controls for a source, the significance of this source should be considered relative to other noise sources on a machine. In this case, the sound level with the electric motor and hydraulics operating was 85 dB(A) whereas noise due to drilling and bolting is about 100 dB(A). In this case, the noise due to the electric motor and hydraulic pumps is insignificant.

Windshields on jumbo drills and bolters reduced the sound level at the operator's station during the drilling/bolting cycle up to 3 dB(A). The noise generated from drilling and bolting is relatively high frequency in nature. Therefore, the windshield provides an effective barrier. Gaps between and around sections of the windshield should be sealed for the most effective noise control. In addition, wrapping a windshield around the operator station improves the noise reduction by forcing drilling/bolting noise to travel further around the windshield to get to the operator. In addition, this helps block noise reflected from the rib from reaching the operator.

6.4 Lessons Learned

Through evaluating different noise controls on underground machinery, NIOSH researchers discovered several findings. Both the effective and ineffective treatments rendered valuable information.

1. Although it is tempting to use sound-absorbing materials for noise controls because they are inexpensive and simple to attach to existing surfaces, sound barriers were always more effective in the examples NIOSH studied for this report.

2. Windshields and environmental cabs can be highly effective noise controls, especially for high frequency noise.

3. Plugging gaps in machine panels and windshields with a material that creates an airtight seal can greatly enhance the noise reduction benefits of existing barriers.

4. Gaps in barriers compromise noise control effectiveness.

5. When openings in enclosures are necessary, a partial enclosure can provide some benefit. Enclosures should be lined with an absorptive material thick enough to absorb the dominant sound frequencies. Openings to let air in and out of the enclosure should have lined ducts with multiple bends to absorb sound and to force it to follow a circuitous pathway before exiting the enclosure.

In order to reduce noise-induced hearing loss from work-related noise exposure, mine workers, union representatives, mine managers, equipment manufacturers, NIOSH, and MSHA must work in partnership to successfully construct and implement new and better noise controls. To ensure the success of a noise control program, appropriate materials must be applied and the noise sources treated must be significant in terms of the worker's daily noise exposure.

References

ANSI [2001]. American national standard: measurement of occupational noise exposure. New York: American National Standards Institute, Inc., S12.19–1996.

ANSI [2007]. American National Standard Specification for Personal Noise Dosimeters. New York: American National Standards Institute, Inc., S1.25–1991 (R2007).

CFR. Code of Federal Regulations. Washington, DC: U.S. Government Printing Office, Office of the Federal Register.

Cowan JP [1994]. Handbook of environmental acoustics. New York: Van Nostrand Reinhold, p. 37.

Kryter KD [1994]. The handbook of hearing and the effects of noise. San Diego, CA: Academic Press.

Moulder R [1998]. Sound absorptive materials. In: Harris CM, ed. Handbook of acoustical measurements and noise control. 3rd ed. New York: McGraw-Hill, Chapter 30.

NIOSH [1979]. Industrial noise control manual. Cincinnati, OH: U.S. Department of Health, Education, and Welfare, Center for Disease Control, National Institute for Occupational Safety and Health, SHEW (NIOSH) Publication No. 79–117.

NIOSH [1996]. National occupational research agenda. Cincinnati, OH: U.S. Department of Health and Human Services, Centers for Disease Control and Prevention, National Institute for Occupational Safety and Health, DHHS (NIOSH) Publication No. 96–115.

Tak S, Calvert GM [2008]. Hearing difficulty attributable to employment by industry and occupation: An analysis of the National Health Interview Survey—United States, 1997 to 2003. Journal of Occupational and Environmental Medicine 50:46–56.

Tak S, Davis RD, Calvert GM [2009]. Exposure to hazardous workplace noise and use of hearing protection devices among US workers—NHANES, 1999–2004. American Journal of Industrial Medicine 52:358–371.

Appendix A — What is Noise?

Noise is generally defined as sound that is loud, unpleasant, or otherwise undesirable. As a type of sound, it can be characterized according to its physical properties.

A.1 The Physics of Sound

A basic understanding of the physics of sound is crucial for measuring and controlling it. Sound results from a physical occurrence created by a pressure disturbance that travels, or propagates, through an elastic medium via longitudinal wave motion. When this disturbance reaches the ear, it is perceived by the brain as sound. The propagation of sound can be compared with the motion that takes place along a stretched spring. If one end of the spring is repeatedly compressed and released, a traveling wave is produced. As the wave propagates, sections of the spring compress and then expand. This compression and expansion of the spring is analogous to the compression and expansion of air particles caused by the propagation of a sound wave. The compression and expansion of the air particles cause a fluctuation of the air pressure above and below the atmospheric pressure. This is shown graphically in Figure A.1.

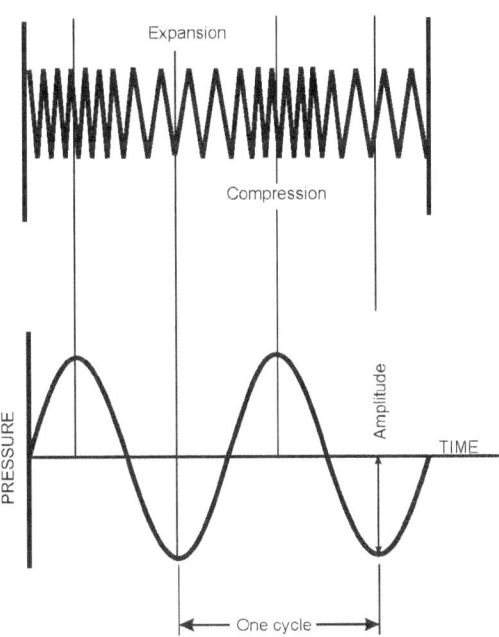

Figure A.1. An illustration of wave motion (sine wave).

Appendix A | What is Noise?

The rate at which the air particles expand and contract is referred to as the frequency of the wave. Frequency is measured in cycles per second and is commonly expressed in Hertz (Hz). Figure A.1 depicts a pure tone that has only one frequency. For a pure tone, the frequency can be determined by counting the number of full cycles that occur in a fixed time and then dividing the number by the time. For example, Figure A.1 shows two full cycles. Assuming the two cycles are completed in 0.01 seconds, the frequency would be 200 cycles per second, or 200 Hz. Young listeners with normal hearing can usually hear sounds from a low frequency of about 20 Hz to a high frequency of roughly 20,000 Hz.

In general, real sounds are made up of many frequencies. Knowing the frequency content of a sound is useful in determining how to design noise controls to attenuate offensive sounds, or noise. The frequency content of a particular noise can be used to determine which materials to use to control it. In addition, the frequency content of a noise can be used to determine its source.

The time it takes for the wave to go through one complete cycle, from trough-to-trough (minimum-to-minimum) or crest-to-crest (maximum-to-maximum), as shown in Figure A.1, is the period of the wave. The period (T) is inversely related to frequency (f) by:

$$T = 1/f \qquad (A.1)$$

The distance a sound wave travels during one complete cycle is its wavelength. Wavelength is a distance commonly expressed in meters or feet. The wavelength of a sound, λ, can be found using the frequency, f, and the speed of sound, c by:

$$\lambda = c/f \qquad (A.2)$$

The speed that a sound wave travels in any medium depends only on the properties of the medium through which it propagates. The speed of sound in dry air at a temperature of 20 °C, or 68 °F, is about 343 meters per second or about 1,125 feet per second. Using equation A.2, the wavelength for a 100 Hz tone in air would be 3.43 meters, or 11.25 feet.

Assuming the speed of sound is held constant, Equation A.2 shows that as frequency increases, wavelength must decrease, and vice versa. For high frequency sounds, the distance between each area of compression or expansion along the wave will be small. For low frequency sounds, the distance between each area of compression or expansion will be relatively large. Stated another way, low frequency sounds have long wavelengths and high frequency sounds have short wavelengths.

In air, sound travels as a pressure wave. As a pressure disturbance passes a point in space, minute changes occur in the atmospheric pressure at that point. The sound pressure alternates between positive and negative values so it adds to and subtracts from the atmospheric pressure. This sound pressure is analogous to an AC voltage on top of a DC voltage. Here, atmospheric pressure is analogous to the DC voltage.

Sound pressure is typically measured in Pascal (Pa). One Pascal is equal to 1.45×10^{-4} pounds per square inch (PSI). Sound pressure causes the eardrum to vibrate. These minute changes in pressure can be measured with a microphone. The magnitude of the pressure change corresponds to the amplitude of the wave as shown in Figure A.1. The perceived loudness of a sound is directly related to the amplitude of the disturbance. Several other factors, such as frequency and fluctuations in the amplitude of a sound, also influence the perceived loudness of a sound.

To quantify the change in the pressure at any point due to a sound wave, the root-mean-square (RMS) pressure is used. The RMS is the square root of the average of the instantaneous pressure squared. If a simple average of the changes in pressure were used to quantify the pressure wave, the result would be equal to zero because the positive pressure fluctuations would be cancelled by the negative pressure fluctuations.

The human ear responds to a wide range of pressures from 20×10^{-6} to 20 Pa. The threshold of hearing for an undamaged ear corresponds to an RMS pressure of 20×10^{-6} Pa at 1,000 Hz. An RMS sound pressure near the high end of the range would be painful and could damage the auditory system, even with brief exposures. Working with values of pressure that cover such a large range can be unwieldy. Therefore, to simplify the numbers and reduce the range of values to a manageable size, the value of the RMS sound pressure is converted to decibels (dB).

A.2 Decibels

Decibels are logarithmic values formed by taking the ratio of a power to a reference power. In terms of time-varying signals, such as AC voltage or sound pressure, the power is related to the RMS value squared. The reference used when dealing with sound pressure is 20×10^{-6} Pa, which, as mentioned previously, is the RMS pressure that is barely audible at 1,000 Hz. When measured, this pressure would yield a value of 0 dB. The term *level* is commonly used to designate a logarithmic ratio of relevant parameters. The sound pressure level for any sound can be calculated using the following equation:

$$SPL = 10\log \frac{P_{RMS}^2}{P_{ref}^2} = 20\log \frac{P_{RMS}}{P_{ref}} \qquad (A.3)$$

where SPL is the sound pressure level in dB;
P_{RMS} = RMS sound pressure in Pascals; and
P_{ref} = Reference RMS sound pressure, 20×10^{-6} Pa

The RMS sound pressure commonly referred to as the threshold of pain is 20 Pa. Using this value in Equation A.3 yields a sound pressure level of 120 dB.

A.3 Average Sound Levels

Often, one is more interested in determining the average sound pressure level over a period of time rather than the sound level at a particular time. Two commonly used averages are the equivalent continuous sound level, denoted L_{eq}, and the average sound level, denoted L_{avg}.

The L_{eq} is the constant sound level that has the same energy as a fluctuating sound over the measurement time. For example, if the fluctuating sound in the cab of a haul truck produced an L_{eq} of 80 dB over a 1-hour period, the sound would have the same sound energy as a continuous sound of 80 dB over the same 1-hour period. A 3-dB increase in the L_{eq} corresponds to a doubling of the sound energy over the measurement time.

The L_{avg} is commonly used in the assessment of worker noise exposure. In some texts L_{avg} is called TWA or TWA(x) where the 'x' denotes the measurement time in hours. The L_{avg} is similar to the L_{eq} except the value of the L_{avg} depends on the exchange rate, threshold level, and criterion level used to measure a worker's noise dose. For a given criterion level and allowable exposure time, the exchange rate is the change in sound level corresponding to a doubling or halving of the allowable exposure time. On many dosimeters, the exchange rate can be set to 3, 4, 5, or 6 dB. The current MSHA exchange rate is 5 dB.

The 8-hour time-weighted average sound level (TWA_8) is used in occupational noise measurements. It is the constant A-weighted sound level for an 8-hour time period that would expose a person to the same noise dose as did the actual time-varying sound level over the time used to perform the measurement. If the measurement time is less than 8 hours, the TWA_8 will always be less than L_{avg}. If the measurement period is greater than 8 hours, the TWA_8 will always be greater than L_{avg}.

Appendix A What is Noise?

Table A.1 lists various measurement times and their associated TWA_8 values for an L_{avg} of 90 dB(A) using a 5-dB exchange rate. For a measurement time of 4 hours, the corresponding TWA_8 would be 85 dB(A). Further, a measurement time of 12 hours would result in a TWA_8 of about 93 dB and a 16-hour measurement time would yield a TWA_8 of 95 dB.

Assessing an operator's TWA_8 is a very good method to compare noise exposures before and after installing controls. A reduction in TWA_8 that reaches the intended target after implementation of engineering noise controls would indicate that the effort was successful and achieved the desired result. MSHA PIB 08-12 considers a noise control to be technologically achievable if it reduces a worker's noise exposure by 3 dB(A). TWA_8 measurements can also be used to assess compliance with MSHA regulatory requirements.

Table A.1. TWA_8 as a function of Measurement Time for an L_{avg} of 90 dB(A)

Time (hours)	TWA_8 (dB[A])
1	75
2	80
4	85
8	90
10	92
12	93
14	94
16	95

Appendix B: Noise Exposure as a Health Hazard

Noise is "audible acoustic energy (or sound) that is unwanted because it has adverse auditory and nonauditory physiological or psychological effects on people" [Kryter 1994]. Generally, noise is unwanted sound.

B.1 Effects of Noise

As stated above, noise can have both psychological and physiological effects. The psychological effects of noise include annoyance, speech interference, sleep interference, and decreased work performance [Kryter 1994]. Sudden noises can trigger the physiological response of muscular reflex as the body prepares for defensive action against the source of the noise. Sometimes, this spontaneous reaction can interfere with tasks or cause accidents. However, the physiological effect most commonly associated with noise is temporary or permanent hearing loss.

It is widely known that exposure to high sound levels can cause damage to the mechanisms of the inner ear and reduce hearing sensitivity. A noise-exposed individual may experience a temporary threshold shift, which is a temporary reduction in sensitivity to sound. Normal sensitivity may return after a period without noise exposure. However, continued exposure to high sound levels can eventually result in a permanent threshold shift—or noise-induced hearing loss (NIHL). Because NIHL generally occurs gradually over time, it can go unnoticed until a considerable hearing loss has occurred. In some cases, an exposed individual will claim to have become accustomed to the noise. However, in reality, it is impossible for the ears to become resistant to noise.

B.2 Dose Accumulation

In many mining environments, sound levels vary throughout the day. Changes in geology, the amount of material mined, or mining method can all affect the generated sound levels. Also, many miners move throughout the mine or operate their machines in various modes, which can expose the miner to different sound levels. When sound levels are not constant, the total dose for a worker can be calculated using the partial dose for each task or sound level. With MSHA PEL criteria, exposure to any sound level at or above 90 dB(A) results in the accumulation of a noise dose. The total, or daily, noise dose can be calculated from the partial dose due to each task using the following equation:

Appendix B | Noise Exposure as Health Hazard

$$D = \frac{C_1}{T_1} + \frac{C_2}{T_2} + \frac{C_3}{T_3} + \cdots + \frac{C_n}{T_n} \quad (B.1)$$

where D = total noise dose
C_n = the actual exposure time for task or sound level n, in hours
T_n = the allowable exposure time for task or sound level n, in hours (see Table A.1)

If the allowable exposure time is something other than the values listed in Table A.1, the value of T_n based on an 8-hour workday can be calculated from the following equation (assuming MSHA PEL parameters):

$$\text{Allowable Exposure Time (hours)} = \frac{8}{2^{0.2(L_A - 90)}} \quad (B.2)$$

where L_A = the A-weighted sound level at the worker's position for a given task

To calculate the allowable exposure time for a criterion level other than 90 dB(A), the desired sound level in dB(A) would replace 90 in the denominator of Equation B.2.

Appendix C — Sound Measurement

In a mining environment, sound levels are usually measured for either compliance or diagnostic purposes. Compliance measurements are made in accordance with Mine Safety and Health Administration (MSHA) regulations to determine if exposures exceed the Permissible Exposure Level (PEL). Diagnostic measurements, on the other hand, are used to help locate and quantify noise sources. Another important difference is that compliance measurements focus on workers in the vicinity of one or more noise-generating machines, whereas diagnostic measurements primarily focus on the machines themselves. This appendix summarizes basic techniques of diagnostic noise measurement as the first step in planning a successful noise control strategy. The appendix also summarizes diagnostic usage of the two most basic and essential instruments available to measure and characterize sound: sound level meters and dosimeters.

C.1 Sound Level Meters

A sound level meter (SLM) is the instrument used to measure the sound level of a noise source. MSHA regulations allow the use of a Type 1 (precision) or Type 2 (general purpose) SLM as designated by the American National Standards Institute (ANSI S1.4). A Type 2 meter has broader performance tolerances and is usually less expensive. According to ANSI S1.4, the expected total error for an SLM measuring steady broadband noise in a reverberant field is approximately +/- 1.5 dB for a Type 1 SLM and +/- 2.3 dB for a Type 2 SLM.

When using an SLM, it is important to set the frequency weighting and the time weighting to the appropriate settings. For measurements related to assessment of noise exposure, the SLM should be set to "A-weighting" and "slow response". A single sine wave (refer to Figure A.1) produces a sound at only one frequency. This type of sound is called a pure tone. Most real-world sounds consist of many frequencies. Therefore, when an overall sound level measurement is performed, a single decibel value is obtained that accounts for the sound energy over a broad range of frequencies. However, the human ear does not respond to all frequencies in the same way. At low sound levels, the ear is most sensitive to sounds with frequencies from 500 to 5,000 Hz. Below 500 Hz, the sensitivity of the ear falls off quite rapidly as frequency decreases. As the sound level increases, the response of the ear "flattens" somewhat with respect to frequency. For high sound levels, the sensitivity of the ear to sounds below 500 Hz does not decrease as rapidly as it does for low level sounds.

Appendix C Noise Control in Underground Metal Mining

Sound measurements are usually conducted to assess human reaction to sound or to determine the potential risk for hearing damage. Since the sensitivity of the ear is a function of both frequency and level, several weighting networks, or filters, were developed to approximate the sensitivity of the ear. These filters simulate the response of the ear under certain conditions. A typical SLM has two different frequency weighting networks, identified as the A-weighting and C-weighting networks. Some meters also have a B-weighting network, which falls between the A and the C networks.

In terms of judging loudness, A-weighting is most appropriate for low level sounds (40 dB), B-weighting is most appropriate for mid level sounds (70 dB), and C-weighting is most appropriate for high level sounds (100 dB). When assessing the potential risk for causing hearing loss, the A-weighing network is used because it is a reasonable estimator of the risk of hearing damage. Figure C.1 shows the frequency response characteristics of the A, B, and C networks. When a weighting filter is applied in determining the level of a sound, the word *pressure* is dropped from the term *sound pressure level*. If an RMS pressure of 20 Pa is measured using the A-weighted network, sound level would be written as 120 dB(A).

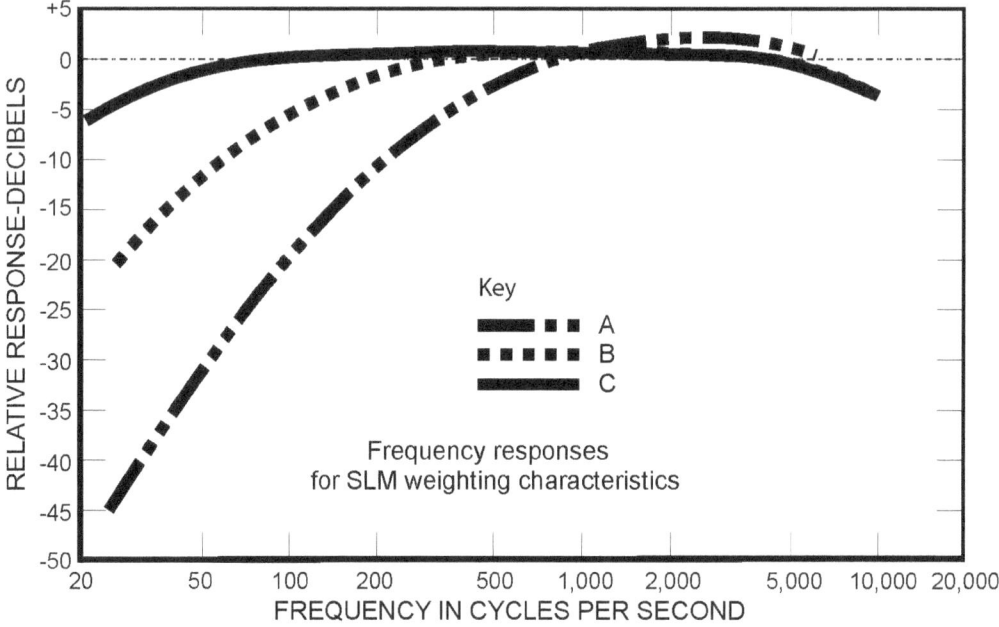

Figure C.1. Figure C.1. Frequency response for weighting networks.

Time weighting, or response, adjusts the response time of the meter. Typically, it can be set to Fast, Slow, Impulse, or Peak. Fast time weighting can be used to closely track rapid fluctuations in sound levels. Slow time weighting is useful when estimating the average level of a rapidly fluctuating sound. For an integrating-averaging SLM, a time weighting of Fast or Slow should not affect the measurement of the equivalent continuous sound level, or Leq. However, for compliance measurements, MSHA requires the use of Slow for the time weighting and A-weighting for the frequency weighting.

Impulse time weighting is used for measuring the sound level of impulsive events such as engine backfires or explosions. This setting is usually not used for measurements related to hearing conservation. Peak time weighting indicates the level of the absolute peak sound level when the level reached its greatest magnitude. This setting is sometimes used to determine the peak level of impulse sounds.

An SLM must be calibrated before and after each measurement period with a calibrator that fits the size of the microphone on the meter. The pre-measurement calibration is necessary to ensure the SLM is functioning properly prior to making measurements. The post-measurement calibration is important in a mining environment because the instrument is likely to be subjected to jolting and jarring during a work shift and because temperature or humidity extremes could affect the accuracy of the meter.

The microphone, the most fragile part of the instrument, is especially susceptible to damage. The documentation provided by the instrument manufacturer should list the operating ranges for humidity, pressure, and temperature. The SLM should be calibrated by a qualified laboratory for a calibration traceable to the National Institute of Standards and Technology at the interval recommended by the manufacturer, typically every 1-2 years.

When performing measurements with an SLM, care should be taken to avoid contaminating the readings with reflected sound. To avoid doing this, the microphone should generally be a minimum of 3 feet, or 1 meter, away from all reflecting surfaces. This includes walls, ribs, ceilings, the ground, and the body of the person doing the measurements. If conditions allow, the SLM should be mounted on a tripod. This will yield a consistent microphone height and further reduce the chances of the measurement being influenced by reflections from the body of the person performing the reading.

After calibrating the SLM, but before measuring the sound of interest, the background sound level, or ambient sound level, should be measured. The background level is the sound level that is measured when the piece of equipment or the device to be measured is shut down. It is desirable to have at least a 10-dB difference between the background level and the sound level measured when the object under test is operating. This is usually not a problem in the mining industry. However, measurements near ventilation fans may have high background sound levels. If a 10-dB or greater difference is not achieved, correction factors must be used to account for the background noise. For example, a measurement taken with a motor operating yielded a result of 87 dB(A) and the level with the motor shut down produced a level of 83 dB(A); the difference is 4 dB(A), which is less than 10 dB(A), and Table C.1 must be consulted. The correction associated with the difference is 2 dB(A). Subtracting this value from the combined sound level of 87 dB(A) yields a result of 85 dB(A).

Air passing over the grid of the microphone, which can happen when using a SLM in a windy environment, is picked up by the microphone element as noise that typically has predominantly low frequency components. This wind noise is a spurious contribution to the actual measurement. Using a windscreen can minimize these contributions. A windscreen is a ball of open-cell foam that is placed over a microphone to minimize wind noise while having minimal effect on the directional and frequency response characteristics of the microphone. The manufacturer of the SLM should be consulted for information regarding proper windscreen selection.

When measuring sound levels, it is important to remember that sound pressure varies with distance from the source. Therefore, for consistent results, it is crucial to record the distance from the noise source to the measurement location. Furthermore, when measuring the sound level of a piece of machinery, the side of the machine from which the measurements are taken should be noted. This is necessary because the sound level around the machine may vary.

Table C.1. Correction for background noise*

Noise level minus background noise level (dB[A])	Correction to subtract from noise level (dB[A])
8 to 10	0.5
6 to 8	1.0
4.5 to 6	1.5
4 to 4.5	2.0
3.5	2.5
3	3.0

*Correction to obtain the sound level of the source if the background noise were not present

C.2 Frequency Bands

The overall level of a sound is an important metric for determining the extent of a noise problem in an occupational setting. However, when a noise problem has been determined to exist, more information is needed to determine how to best approach the problem. An important measure of a noise is its frequency distribution, or spectrum. In some cases, knowing the frequency distribution of a noise will help to identify its source. The spectrum may also help in the selection of the most appropriate noise control materials for a problem.

Octave band analysis is one type of spectrum analysis. An octave is a frequency interval between two sounds with a frequency ratio of 2, such as from 125–250 Hz or 2,000–4,000 Hz. In octave band analysis, the frequency spectrum is divided into bands on a logarithmic basis. Each band is identified by a center frequency, which is the geometric center of the band. The octave band center frequencies commonly used are 63; 125; 250; 500; 1,000; 2,000; 4,000; and 8,000 Hz.

When more detailed information is required, one-third-octave bands are used for spectral analysis. This type of analysis divides each octave band into three logarithmic bands. One-third-octave band analysis is useful for locating the frequency region of tones that may become obscured when using octave band analysis.

C.3 Decibel Addition

To fully evaluate a noise problem, it is usually necessary to sum the sound levels due to multiple noise sources. To calculate the sum of the sources, the sources are assumed to be random and unrelated to each other—or incoherent—as is usual when dealing with machinery or equipment noise.

Since sound levels are formed by taking a logarithmic ratio, simple addition will not yield the correct total sound level. This means that 80 dB + 80 dB ≠ 160 dB. To obtain a correct result of adding two sound levels, first, the difference between them is determined. Table C.2 shows the number of decibels to add to the higher of two sound levels to be added. The corresponding value from Table C.2 is added to the larger of the two original sound levels. For example, in Equation C.1 the two values are the same, so the difference is zero. In this case, 3 dB is added to the higher of the original measurement (in this case they are the same):

$$80 \text{ dB} + 80 \text{ dB} = 83 \text{ dB} \qquad (C.1)$$

If the levels to be summed are 80 dB and 82 dB, Table C.2 indicates that 2.1 dB should be added to the higher level, which is 82 dB:

$$80 \text{ dB} + 82 \text{ dB} = 84.1 \text{ dB} \qquad (C.2)$$

When dealing with decibels, it is generally unnecessary to obtain decimal accuracy. The result of Equation C.2 should be recorded as 84 dB. When adding several decibel levels, it is best to begin by combining the two lower levels and then adding their sum to the next highest level. This trend should be continued until all of the levels are summed and a final result is found.

Table C.2. Chart for calculating the sum of two decibel levels

Difference between two levels being added (dB)	Decibels added to higher level
0	3.0
1	2.6
2	2.1
3	1.8
4	1.4
5	1.2
6	1.0
7	0.8
8	0.6
9	0.5
10	0.4

Source: NIOSH 1979

Delivering on the Nation's promise:
*safety and health at work for all people
through research and prevention*

To receive NIOSH documents or more information about
occupational safety and health topics, contact NIOSH at

1–800–CDC–INFO (1–800–232–4636)
TTY: 1–888–232–6348
E-mail: cdcinfo@cdc.gov

or visit the NIOSH Web site at **www.cdc.gov/niosh.**

For a monthly update on news at NIOSH, subscribe to
NIOSH eNews by visiting **www.cdc.gov/niosh/eNews.**

DHHS (NIOSH) Publication No. 2010–111

SAFER • HEALTHIER • PEOPLE™

www.ingramcontent.com/pod-product-compliance
Lightning Source LLC
Chambersburg PA
CBHW081850170526
45167CB00007B/2961